Mateus Pessanha
Sandra de Andrade

The Dynamics of E-waste Disposal

Mateus Pessanha
Sandra de Andrade

The Dynamics of E-waste Disposal

Management and Sustainability in the Disposal of Electronic Waste in Campos dos Goytacazes - RJ

ScienciaScripts

Imprint

Any brand names and product names mentioned in this book are subject to trademark, brand or patent protection and are trademarks or registered trademarks of their respective holders. The use of brand names, product names, common names, trade names, product descriptions etc. even without a particular marking in this work is in no way to be construed to mean that such names may be regarded as unrestricted in respect of trademark and brand protection legislation and could thus be used by anyone.

Cover image: www.ingimage.com

This book is a translation from the original published under ISBN 978-620-6-76151-8.

Publisher:
Sciencia Scripts
is a trademark of
Dodo Books Indian Ocean Ltd. and OmniScriptum S.R.L publishing group

120 High Road, East Finchley, London, N2 9ED, United Kingdom
Str. Armeneasca 28/1, office 1, Chisinau MD-2012, Republic of Moldova, Europe
Printed at: see last page
ISBN: 978-620-8-03972-1

TABLE OF CONTENTS

ACKNOWLEDGMENTS

The journey through the academic world is not easy at all, so it takes a lot of wisdom to deal with a plural and conflicting world, so I thank God first of all for giving me the wisdom to deal with all the problems, from the personal to the academic.

I would like to thank my family, who gave me all the support I needed to achieve my dream degree. My immense thanks go to my advisor, Sandra Fernandes de Andrade, for her lessons, guidance and for making this research possible. I would also like to extend this thanks, with greater honor, to Glayce Azeredo.

Finally, I am very grateful to have met Bia, Gisele, Fran and Alexia. They were part of this arduous but exciting journey that is the academic world.

SUMMARY

The intensification of production and the use of technology, fostered by the industrial revolutions, has increased access to the consumption of technological products. Thus, with a view to dealing with solid waste, including electronic waste, in 2010 Brazil implemented Law 12.305, which established the national solid waste policy and enabled the creation of municipal policies. In this sense, the municipality of Campos dos Goytacazes instituted its municipal solid waste policy in 2011 by means of Law 8.232. The aim of this study is therefore to analyze the dynamics of dumping and collecting e-waste at the public points specified by the municipality of Campos dos Goytacazes. In order to achieve the proposed objective, a quantitative-qualitative study was carried out, in which national and municipal legislation relating to e-waste was analyzed. Visits were also made to the municipality's collection points and questionnaires were administered via Google forms to the population living in the municipality's main district, using snowball sampling. The work revealed that in the municipality of Campos dos Goytacazes, although there are policies aimed at electronic waste, there is still a need for more publicity about the collection points, as well as their expansion.

Keywords: Programmed obsolescence, consumerism, reverse logistics.

ABSTRACT

The intensification of production and use of technologies, driven by industrial revolutions, has expanded access to the consumption of technological products. To address solid waste, including electronic waste, Brazil implemented Law 12.305 in 2010, which established the National Solid Waste Policy and enabled the creation of municipal policies. In this context, the Municipality of Campos dos Goytacazes established its municipal solid waste policy in 2011 through Law 8.232. This study aims to analyze the dynamics of disposal and collection of e-waste at public points specified by the municipality of Campos dos Goytacazes. To achieve this objective, a quantitative and qualitative research was conducted, analyzing national and municipal legislation related to e-waste. Visits were made to the collection points in the municipality, and questionnaires were administered via Google Forms to residents of the municipality's central district, using the snowball sampling method. The study revealed that, despite the existence of policies aimed at electronic waste in the Municipality of Campos dos Goytacazes, there is still a need for greater dissemination of information about collection points, as well as their expansion.

Keywords: Planned obsolescence, consumerism, reverse logistics.

LIST OF ABBREVIATIONS AND ACRONYMS

ABINEE - Brazilian Electrical and Electronics Industry Association

ARPAnet - Advanced Research Projects Agency Network

CEA - Center for Environmental Education

CONAMA - National Environment Council

E-waste - Electronic **waste**

ENIAC - Electronic Numerical Integrator Computer

GPS - Global Positioning System

WHO - World Health Organization

UN - United Nations Organization

PEVs - Voluntary drop-off points

PMGIRS - Municipal Integrated Solid Waste Management Plan

PNRS - National Solid Waste Policy

WEEE - Waste Electrical and Electronic Equipment

SINIR - National Solid Waste Management Information System

SULIMP - Public Cleaning Superintendence

UENF - State University of Northern Rio de Janeiro

INTRODUCTION

The existence of today's electronics was only possible because of the advances brought about by the First Industrial Revolution, the main milestone of which, according to Canêdo (1986), was the use of machines to replace part of human manual labor and the social and economic transformation brought about in England in the 18th century.

Through production models such as Fordism and later Toyotism, there was an intensification in the use of technologies, as well as the precariousness of work. In this way, while technologies are expanded, they are also under the control of developed countries, which, from the 1990s onwards, reached a peak in the disposal of solid waste, including electronic waste (e-waste), in other territories. Because of this, they created the Basel Convention, which is part of the process of preventing the transportation and disposal of hazardous waste, in this case prohibiting its disposal in other territories without consent (UN, 2012).

Therefore, according to Pasquini (2020), the 20th century was marked by technological development, because we have the integration between technique and science, which makes it possible to expand medicine, industry and, above all, the emergence of the internet, computers, cell phones, among others. However, we understand that technology can't just be seen as the solution to all of society's problems, because it can also cause some complications, such as e-waste.

We should therefore add that, as technologies get closer to society, we see a reduction in the lifespan of products, as there will be no obstacles to capitalism; likewise, we get closer to consumerism and are caught up in a web of buying, discarding and buying.

Thus, with regard to e-waste, which is the subject of this research, Toyotism has essential characteristics for elucidating concepts such as perceived obsolescence, programmed obsolescence and reverse logistics, since the Toyotist production model, unlike Fordism, which brings mass production and consumption, seeks *just-in-time* production (products follow demand, production has a certain time and the quantity must be exact). Based on the above, in the Fordist production model, the durability of products was greater and consumption had to be constant, however, these characteristics were crucial to its crisis, i.e. the accumulation of products and low demand. Therefore, Waste Electrical and Electronic Equipment (WEEE), if thought of in the Toyotist era, is produced on demand and is less durable.

Therefore, we understand that in the current century (21st), in which we are experiencing a huge consumption of technology and the consumption of technological equipment, such as smartphones, we question whether the laws on electronic waste are really efficient, whether the collection points really work. Consequently, we question who reverse logistics is for and which social agents benefit the most from the imposition of perceived and planned obsolescence, consumerism and, finally, what will be the function of the soil as the foundation of human needs, provider of wealth and support for the toxicity of capitalist production, the aggressors of the environment.

1.1 Objectives

The general objective of this research is to analyze the dynamics of dumping and collecting electronic waste at the public points specified by the municipality of Campos dos Goytacazes, located in the interior of the state of Rio de Janeiro. With regard to the specific objectives, this work will:

1) Identify the existing collection points from Campos dos Goytacazes City Hall;
2) Verify, through the agents of the collection points, the forms of treatment of waste electrical and electronic equipment disposed of at these sites;
3) Detect the companies located in Campos dos Goytacazes that collect e-waste.

1.2 Justification

This work is justified because, when asked what waste is, the central idea that emerges is related to consuming and discarding, but in the case of e-waste, there is often a lack of understanding of recycling processes, seen as something related only to materials such as plastic and cardboard/paper. In addition, WEEE, like plastic, exists in large quantities, so it is not yet clearly exposed to consumerist society as potentially toxic to the soil and harmful to the lives of those who live in places where these materials are disposed of illegally or due to a lack of knowledge of specific disposal sites (Magera, 2012).

Based on the above, electronic objects bring with them three aspects: firstly, the process of technologization of human society (unevenly brought about); secondly, advertising and consumerism; and thirdly, disposal and recycling (Magera, 2012). Thus, when we refer to e-waste, we question why a cell phone, among other equipment, becomes bad or unusable, when it comes to useful life, in a short period of time. In addition, we asked how e-waste should be disposed of, i.e. whether it should be disposed of in the regular garbage or, if available, at specific collection points for this material. Finally, we seek to understand and analyze e-waste beyond consumption and use, in turn seeking to see it when it no longer has any use and its only purpose is disposal.

1.3 Methodology

This work has as its theoretical support works such as Marx; Engels, 2020) and (Canêdo, 1986), which explain the first Industrial Revolution and its consequent technological development in an intelligible way. With regard to the description of e-waste, (Magera, 2012) plus the (Global e-Waste Monitor, 2020) will be important for an in-depth description of the topic. In this way, we have included other renowned references such as (Santos, 1996) and (Santos, 2001), whose name should be exposed in addition to the reference, the eminent Milton Santos, who regardless of the method used, cannot fail to be cited, especially for future geographers.

In order to achieve the proposed objectives, this research was carried out using a quantitative and qualitative approach, also known as a mixed method. According to Minayo (2011, p. 22), the qualitative and quantitative approaches "are not incompatible, [...] between them there is a complementary opposition which, when well worked out theoretically and practically, produces a wealth of information, depth and greater interpretative reliability". Therefore, with the municipality of Campos dos Goytacazes as the *locus* of investigation, data on the municipality was initially collected through the National Solid Waste Management Information System (SINIR), on the municipality's website (in the area relating to solid waste disposal and collection) and, subsequently, analysis was carried out of Federal Law No. 12.305, of August 2, 2010 and Municipal Law No. 8.232, of June 15, 2011, as well as Resolution 401/2008 of the National Environment Council (CONAMA).

After collecting and analyzing documents, visits were made to collection points and to the Municipal Department of Urban Planning, Mobility and the Environment, where data was obtained on the amount of e-waste disposed of and whether there are other collection points. In addition, questionnaires were administered via the Google Forms platform, asking questions about e-waste, disposal and reverse logistics. The questionnaires were structured in a mixed way, i.e. there is a part where the questions cannot be answered in a generalized way, i.e. the multiple-choice questions, and another part that can be answered in a generalized way, the written questions. In addition to the above, the questionnaires were applied using snowball sampling, which is described by Vinuto (2014, p. 203) as:

> Snowball sampling is a form of non-probability sampling that uses reference chains. In other words, from this specific type of sampling it is not possible to determine the probability of selecting each participant in the survey, but it is useful for studying certain groups that are difficult to access.

Finally, this type of sampling aims to reach, from specific people, other people who will answer the questionnaire, i.e. groups that would be difficult to reach. Thus, Vinuto (2014) says that snowball sampling starts from the idea of key informants, who will be the basis for locating people with the research profile and, after that, the (key) people selected will first indicate other people from their personal network, thus expanding the application of the questionnaires.

Figure 1 below shows the flowchart that represents the stages described in this methodology. In summary, we start with the initial stage (data collection), the second stage, which is related to fieldwork (visits to the e-waste collection points), the third stage is linked to the preparation and application of the questionnaires, and the last stage is related to the preparation and organization of the data after applying the questionnaire.

Figure 1 - Flowchart of methodological steps

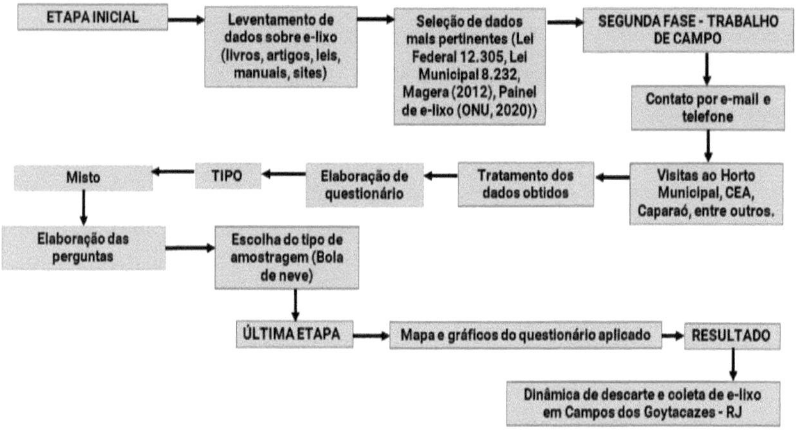

Source: Own authorship.

CHAPTER 1: The industrial revolutions and the process of technological evolution

Before the advent of the First Industrial Revolution, there were traces of attempts to make life easier for human society at the time when it came to building machines. In this case, looking at the Renaissance period, we can mention the eminent Leonardo da Vinci, who, as well as being a painter and anatomist, was also an inventor. Da Vinci is the owner of the Mona Lisa, the Vitruvian Man, the Flying Machine and so many other inventions that can be considered machines, such as the Aerial Screw (a glimpse of an aerial vehicle like the helicopter). However, we can't limit inventions - many of which were only recorded in drawings - such as Da Vinci's, or others that were put into practice, such as the appearance in 1608 of the instrument for observing objects from a distance, developed by Hans Lippershey and further developed by Galileo Galilei, which gave rise to the astronomical telescope in 1609, as the breaking point in a way of living, acting, that is, working and accumulating capital. Thus, in the First Industrial Revolution we have the necessary foundations for understanding the use of machines and their expansion in the ensuing revolutions.

This chapter provides a brief analysis of the industrial revolutions and how machines have transformed the world of production and society. In this way, we will try to explain the differences between tools and machines and highlight, in another aspect, the transformation of capitalism based on the Fordist and Toyotist production models. In addition, there will be theoretical and explanatory support through the work History, Nature, Labor and Education (Marx; Engels, 2020), which exposes the relationship between merchandise, technology and the industrial revolution, as well as The Industrial Revolution (Canêdo, 1986), which will be the support for the first work.

1.1 The First Industrial Revolution

The First Industrial Revolution began in England around the middle of the 18th century. As Chiavenato (2003) explains, England's pioneering spirit was linked to its naval supremacy, the availability of natural resources, the availability of labor, the availability of capital and the parliamentary monarchy, which was important for the fall of absolutism and the rise of the bourgeoisie in other sectors such as politics. Therefore, with all the intersecting bases, the First Revolution can be analyzed as the transition "[...] from a society with an agrarian and artisanal economy to another dominated by industry and machinery" (Canêdo, 1986, p. 6).

The First Industrial Revolution saw the machine as the protagonist. Thus, if we look at it in terms of technological innovations, Canêdo (1986) gives a summary consisting of three innovations, the first of which is linked to the emergence of faster and more precise machines, replacing manual labor; the second refers to the use of steam to drive the machine, replacing other energies such as muscle, wind and water; the last innovation is the improvement in obtaining raw materials, such as minerals, which were important drivers of metallurgy and the chemical industry. We should also point out, following the path of these innovations, that in this period there was no consonance between technique and science. This is also corroborated by Canêdo (1986, place page), who says that "improvements in work tools were the result of technical discoveries made at random, with no connection to rational research applied to practice".

On the other hand, we observe the distinction between machine and tool, the former being characterized as "[...] a natural force other than human, such as animal, hydraulic, wind power, etc." and the latter, the tool, "[...] man would be the driving force" (Marx; Engels, 2020, p. 366). In this context, with the expansion of tools to machines, for example, from the handloom to the mechanical loom, we have what Marx and Engels (2020, p. 368) call a machine tool, which would be "[...] a mechanism which, when the corresponding movement is transmitted to it, performs with its tools the same operations that the worker performed before with similar tools". Having said that, we can say, following Braverman (1980), that machines are evolving and mechanical processes are increasingly becoming automatic, reducing human intervention.

From the above, we look at the organization of productive work, which ceases to be, as Canêdo (1986, p. 6) states, "[...] home production of the article that served a small market" [...] and becomes characterized by the "[...] existence of factories equipped with steam-powered machines, grouping up to hundreds of workers engaged in manufacturing [...]".

Thus, in the First Revolution, consumption was concentrated on machines, tools, bridges, pipes, building materials and household utensils (pots, knives, etc.), due to iron, Hobsbawm (2000). In this sense, Hobsbawm (2000) also adds the creation of factories for the production of furniture and clothing, which in turn generated the consumption of shoes, dresses, shirts, pants, closets, tables, among others.

1.2 Second Industrial Revolution: from Taylorism to Fordism, from Fordism to Flexible Accumulation

Following the Second Industrial Revolution, we are faced with the Fordist production model (which began in 1914), which has its roots in the precepts of Scientific Management developed by Frederick Winslow Taylor. According to Ribeiro (2015), this management created rules and standardized ways of carrying out work through methods of experimentation. These rules would be obtained by the relationship between time and movement, mediated by standardized methods. Thus, Braverman (1980, p. 82) says that "scientific management, as it is called, means a commitment to apply the methods of science to the complex and growing problems of controlling work and production in rapidly expanding capitalist enterprises".

Later on, when Henry Ford took advantage of Taylor and applied his principles to the conveyor belt and the proletarian, just like Chaplin in Modern Times, we see, taking into account the principles mentioned in table 1 below, the distancing of the worker from production, becoming, in turn, just another arm of the machine, unaware of the technical part, which is retained in management, as well as the distancing from the design phase, thus appearing to be just a puppet.

Table 1: Taylor's Principle of Scientific Management

Taylor's Principles of Scientific Management	
1st Principle	This principle is seen as dissociating the work process from the workers' specialties, i.e. the work process must be independent of the workers' trade, tradition and knowledge. From then on, it must depend not entirely on the ability of the workers, but entirely on managerial policies
2nd Principle	This is the principle of the separation of design and execution, rather than its more common name of the principle of the separation of mental and manual labor. Therefore, conception (mental) and execution (practical) must become separate spheres of work, and to this end the study of work processes must be reserved for management and prevented from being carried out by workers, to whom its results are communicated only in the form of simplified functions, guided by simplified instructions which it is their duty to follow without thinking and without understanding the technical reasoning or underlying data
3rd Principle	In this principle, the essential element is the pre-planning and pre-calculation of all the elements of the work process, which no longer exists as a process in the imagination of the worker, but only as a process in the mind of a special management team. Finally, there is the monopoly of knowledge to control every phase of the work process and how it is carried out.

Source: Braverman (1980, p. 103-108).

On another point, even though Ford used Taylor's principles, Ribeiro (2015, p. 71) highlights the difference between Taylorism and Fordism, the latter differing from the former because it sought "not [...] just to dominate the workforce, it wanted to win their adherence". In view of this, the main characteristic of the Fordist model is mass production for mass consumption, i.e. the mass production of durable goods.

In this sense, at the end of the 1960s, Fordism went into crisis, generating, as Ribeiro (2015, p. 73) puts it, "[...] a saturation of the market, [...] which leads (to) a decreasing rate of consumption of these goods". In this way, we can add that the crisis of Fordism was also influenced by the oil crisis that occurred in the same period due to the protectionism of Middle Eastern countries when they learned that oil is not a renewable commodity, reducing its production and ultimately causing the price of a barrel to rise (Harvey, 1992). In the period following the height of the crisis, we should point out that there was economic, social and productive restructuring. In view of this, Harvey (1992, p. 140) exposes some of the factors evident in the period of uncertainty of this restructuring, i.e., "[...] a series of new experiments in the fields of industrial organization and social and political life began to take shape", then, we observe "[...] the first signs of the transition to an entirely new regime of accumulation, associated with a very distinct system of political and social regulation". We therefore have the emergence of flexible accumulation, whose structure is sustained by the flexibilization of "[...] work processes, labour markets, products and consumption patterns" (Harvey, 1992, p. 140).

Toyotism, called flexible accumulation by Harvey (1992), is a form of work organization that emerged with Toyota, a Japanese car manufacturer. Unlike Fordism, the Toyotist production model gives workers new ways of carrying out their work and the knowledge they must possess. In this way, moving away from the Fordist production model, which did not require in-depth technical knowledge and whose tasks were carried out by the performance of a single function, Toyotism generates a worker who performs several functions within a factory, such as, for example, assembling a car tire, manipulating some machine for transporting other parts and, therefore, must possess technical knowledge. In the same vein, Abramides and Cabral (2003, p. 4-5) say that this production model consists of "productive restructuring (based on) increased productivity, efficiency, quality, new forms of technology and management, made effective through technological innovations." This can be seen in the table below, which sets out issues related to the outsourcing of work, meritocracy (which generates competition within workplaces) and production at the right time and in the right quantity, which in turn reduces the accumulation of products.

Dimensions of Toyotism	
1st Dimension	The employment system adopted by large companies consists of: lifetime employment (although there is no formal contract, promotion for length of service, the worker is not hired for a job, but for the company, in a specific position, with a salary.
2nd Dimension	Work organization and management system: just-in-time - producing at the right time, in the right quantity; Kanban - control cards or passwords for replacing parts and stocks; total quality - involving workers to improve production; teamwork - work organization is based on groups of multi-skilled workers who perform multiple functions.
3rd Dimension	The union representation system: company unions are integrated into labor management policy. The positions held in the company are merged with those of the union
4th Dimension	System of inter-company relations: very hierarchical relations between large companies and small and medium-sized ones. Small companies are subcontracted to extremely precarious and unstable micro-enterprises.

Source: Drunk 1999 apud Ribeiro, 2015, p. 74

As we explained earlier, one of the reasons for the crisis of Fordism was the decreasing consumption of products, which led to their accumulation. Therefore, when the characteristics of Toyotism, highlighted in Table 2, are incorporated into Fordism, we see that durability, i.e. the shelf life of products, is no longer a problem, since in the Toyotist production model products have a shorter shelf life, are made on demand and have several models. From the above, we have the concept of programmed obsolescence and perceived obsolescence, which will be elucidated in the next chapter and are intertwined in the dynamics of e-waste disposal.

Now, returning to the Second Industrial Revolution, unlike the first revolution, it was more evident in the United States, with Fordism as a milestone, and in this sense we have the use of oil and the discovery of electricity and steel replacing iron (expansion of chemistry). There was also greater proximity between technology and science, as well as the presence of metallurgy, steelmaking, automobiles and, therefore, the emergence of the telegraph and railroads, etc.

Thus, during the middle of the 20th century, industries were "[...] based on high technology, high-tech" (Pasquini, 2020, p. 33). In this context, there is an integration between technology and science, leading us to technological capitalism, which, through technology, brings countries closer together in terms of their cultures, technologies and economies, giving rise to what we call globalization, which became more evident in the 1990s. That said, because of globalization, a concept that will be explained in the next chapter, we are directed towards consumerism and environmental issues such as waste disposal and its harmful side if it is deposited untreated on the ground. Finally, from here on in, we'll begin the journey of electronics beyond the joy and ease, in other words, electronic waste and its problems.

CHAPTER 2: E-waste: a characterization from the technical-scientific-informational environment

The Second World War (1939-1945) marked the entire planet with the atrocities committed by Hitler, the main Nazi leader of the time, and the atomic bombs dropped by the United States on Hiroshima and Nagasaki, Okuno (2015). On the other hand, if we think about technological innovations, there have been advances (good and bad) in terms of the flexibility to move and communicate over land and sea. According to Okuno (2015), one of the (bad) technological and scientific advances was the nuclear bomb itself, such as Little Boy and Fat Man, dropped on Japan. On the other hand, there were advances in aeronautics and, above all, in the calculation of ballistic trajectories, including the Electronic Numerical Integrator Computer (ENIAC), which began to be developed (by Alan Turing) in 1943 and was only completed and launched in 1946.

In this way, we can see that the aforementioned advances led to the expansion of technologies. As a result of this post-war scenario and the start of the Cold War, advances were even more profound in the area of communications, including, albeit restrictedly, the creation in 1969 of the first computer network (for military purposes), the Advanced Research Projects Agency Network (ARPAnet), as well as, in the 1970s, the launch of the Global Positioning System (GPS), the launch of computers (Apple I, IBM, etc.) and cell phones (Motorola Dynatac 8000X, the first cell phone to be made available for sale). Generally speaking, technologies came to solve humanity's problems, shorten the communication path, expand the development of science in medicine, industry and entertainment. However, during the 1990s, with rapid globalization, more advanced cell phones, computers and televisions began to be marketed without restrictions, which led to the growing consumption of these products.

Therefore, this chapter takes a different perspective on electronic products, because we can't exalt the capitalist view of consumption, in which we just consume and avoid the problems related to the extraction of raw materials, the workers behind the product and, finally, the fate of everything we buy. It is also worth highlighting the reduction in the durability of electronic equipment, clothing and cars, in other words, programmed and perceived obsolescence, which is everywhere and is part of yet another cycle of capitalism, which seeks to commodify everything from nature to bodies.

Without further ado, the basic works of this chapter will be For Another Globalization (Santos, 2001), which exposes important aspects of globalization, such as understanding it as a fable, perverse or possible; The Paths of Garbage (Magera, 2012), an important work for the analysis of the dialectical relationship between garbage and society, the Global Electronic Waste Monitor (UN, 2020) and finally, Society of the Spectacle (Debord, 1967) will be used as a supporting work.

2.1 Technical-scientific-informational environment

When we refer to technologies, we are corroborating their character of transforming and bringing places closer together, in other words, they transform geographical space and, in turn, reduce space-time (Harvey, 1992).

Therefore, based on technological advances, we have the definition of the technical-scientific-informational environment, which for Santos (1996, p. 70) is "[...] the historical moment in which the construction or reconstruction of space will take place with an increasing content of science and technique". In this way, we can see that there is an artificialization of geographical space and, as a result, we are directed towards globalization.

In light of the above, globalization can be understood as economic, technological and cultural integration, in other words, "it is the culmination of the process of internationalization of the capitalist world" (Santos, 2001, p. 23). In addition, Santos (2001) presents three classifications of globalization: globalization as fable, perversity and possibility. In the first classification, that of fable, he relates it as a shortening of distances, with the world appearing to be in the palm of one's hand, this would be a reflection of a market that would be presented as capable of homogenizing the world, but local differences are deepening more and more; in the second, perversity, in which unemployment becomes chronic, poverty and hunger increase, the middle classes are shaken, diseases return, infant deaths remain and education becomes more inaccessible; finally, globalization as a possibility is more humane, it brings about a mixture of peoples, cultures and tastes, it is then a question of the existence of a socio-diversity, much more significant than biodiversity, there is the emergence of a popular culture that uses the technical means previously exclusive to mass culture.

With regard to these classifications, we can draw a parallel with Figures 2 and 3 from the Global e-waste Monitor (2017), which provide information on the number of countries, the total population, the amount of e-waste per inhabitant and the percentage of technological waste generated on the African and American continents. Thus, the African continent represents 5% of the total e-waste generated in the world and 1.9 kg per inhabitant, while the American continent represents 25.3% and 11.6 kg per inhabitant. So, if we make a comparison in relation to the context mentioned in the previous paragraph, is there a globalization of possibility? perverse? or a fable? We can see that there is an unequal globalization and if we go deeper, based on the news about e-waste disposal on the United Nations (UN) website (news.un.org), we come across names like Ghana and Nigeria.

Figure 2 - E-waste in Africa

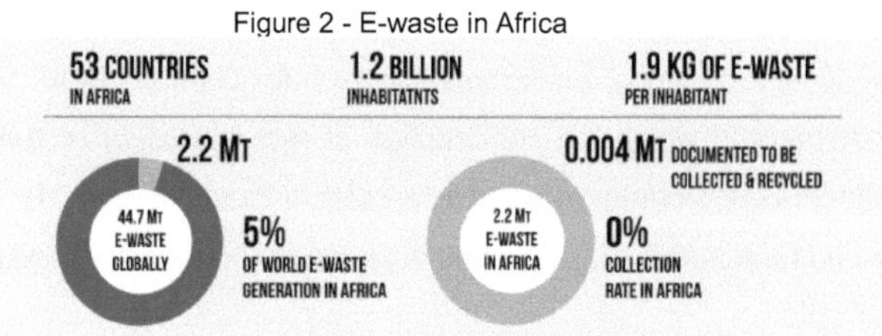

Source: The Global E-waste Monitor (2017, p.61)

Figure 3 - E-waste in the Americas

Source: The Global E-waste Monitor (2017, p.65)

On the other hand, according to figure 4, Asia accounts for most of the generation of electronic waste on Earth. Despite this, it should be noted that some countries in Southeast Asia, such as the Philippines, used to be e-waste dumps. However, since 2019, the country in question has been fighting these illegal dumps, which turn many territories into garbage dumps, when Canada tried to dump waste on its territory. We should also point out that although Asia is one of the biggest generators of e-waste in the world, it has countries like Japan and China, for example, which already deal with their own e-waste. In addition, we note that the amount of waste per person is lower than in America, whose population stands at one billion, which could lead to inequality in terms of technology acquisition.

Figure 4 - E-waste in Asia

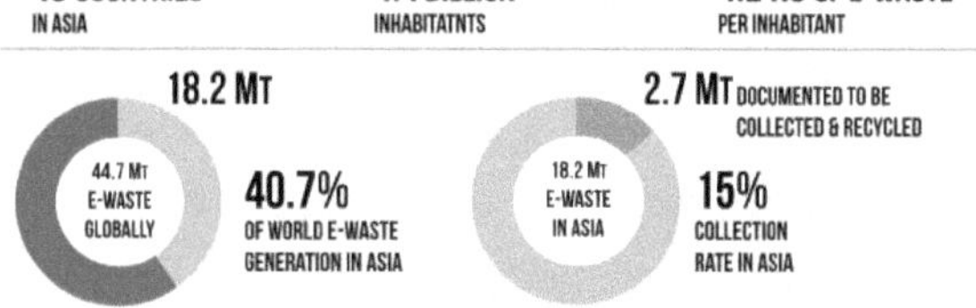

Source: The Global E-waste Monitor (2017, p.69)

In figure 5, which refers to Europe, we have, unlike the continents mentioned, a more extensive charging rate for the management of this waste. As a result, the continent shown is the second largest generator of e-waste in the world and is also second in terms of the amount of technological waste per inhabitant. Extending the analysis to figure 6 below, Oceania, although it generates less waste, has the highest rate of e-waste per person, despite having 13 countries and the smallest population.

Figure 5 - E-waste in Europe

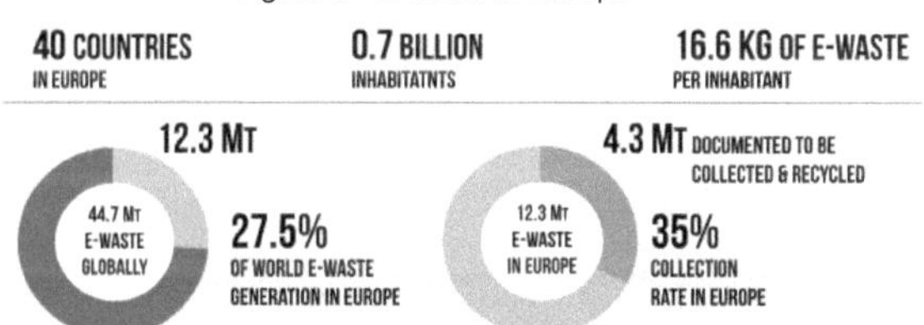

Source: The Global E-waste Monitor (2017, p. 73)

Figure 6 - E-waste in Oceania

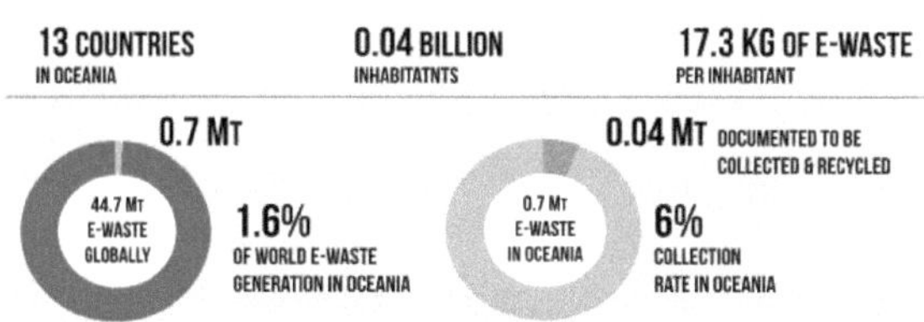

Source: The Global E-wast (2017, p.77)

Therefore, by summarizing all the figures shown, we can see that the process of technological insertion and development is concentrated in Europe, Asia and America. However, we can't just use this data to say that there aren't countries on these continents that are dumping grounds for other so-called developed nations. Therefore, the purpose of these figures is to show, on a smaller scale, the generation of technology around the globe. Finally, the African continent is the one that, in addition to the colonization caused in the previous period, which made many territories on this continent precarious, suffers the most from the disposal of e-waste and a certain international neglect of the problems that waste electrical and electronic equipment (WEEE) can generate in those who handle these products without their integrity or original composition.

2.2 Electronic waste, planned and perceived obsolescence and consumerism

We can consider electronic waste as technological products (cell phones, cameras, televisions, printers, among others) that have perished and no longer perform any function (of reuse) for their owners. As a result, these products are often discarded on a global scale in underprivileged countries whose livelihood, as in Ghana, is the collection and subsequent sale of the most valuable parts. Returning to a larger scale, directed at the object of study of this research, many places, in turn, use legislation at municipal, state and federal levels to promote the creation of collection points for materials that cannot be disposed of in the common garbage, as is the case of the municipality analyzed, Campos dos Goytacazes - RJ.

From the above, WEEE is related to toyotism, and as Magera (2012) reports, in this production model, the responsibility for quality production is retained by the worker, resulting in control by targets through the technique, which is part of the cost of production, i.e. the proletarian who manages to provide models and methods at a lower cost and which still make quality production possible, will be rewarded. This quality production with cheaper methods also reduces the lifespan of the products. Therefore, programmed (planned) obsolescence emerges, which is related to the shelf life of goods, i.e. so that there is no accumulation of products, in toyotism production does not seek to stockpile, but to go according to demand and with the shortened life of products, a new purchase occurs. In addition, the toyotist production model imposes different classifications on a type of product, for example, a cell phone from a brand called X was launched in January with a blue color, however, the following month, the brand launched the same cell phone, but with an orange color and a drawing of a sun, it can be seen that there were large investments in advertising for the cell phone launched in February, finally, a supposed buyer called M, who bought the blue X cell phone, wanted to buy the orange X cell phone with the sun because his would be outdated. In the example above, we have another concept of obsolescence, which is perceived obsolescence, i.e. the product is not bad, but it is outdated in appearance (in fashion).

Thus, according to Magera (2012), the concept of planned obsolescence arose in 1932 with the American real estate investor Bernard London, whose theory consisted of reducing the life cycle of products and, consequently, consumers would return to shopping, so this reduction would also generate more demand for products and there would be more work. In this case, we are referring to the 1929 crisis (financial crisis in the United States), because, according to Magera (2012), products at that time, due to technical and scientific advances, had a longer shelf life, and as the American population, due to the crisis, kept products for longer, this led to the crisis being prolonged. Therefore, looking back to the 1990s, a period when globalization was in evidence, we see that:

> Programmed obsolescence and e-waste are intensifying [...], forcing many countries to create laws to protect their imports, in other words, e-waste is a 'child' that nobody wants to keep. However, in this new model of capital, the poor countries of Asia and Africa end up receiving a large part of the international waste (Magera, 2012, p. 45-46).

Since many countries, especially those in Europe and North America, have turned some countries in Africa and Asia into their dumping grounds, the Basel Convention on the Control of Transboundary Movements of Hazardous Wastes and Their Disposal was created in 1989, but only came into force in 1992. This convention aims to recognize the sovereignty of countries and inhibit the disposal of hazardous waste in their territories in a way that is disrespectful to the population and harmful to the environment. Despite this, it serves as a form of consent, i.e. if country A wants to dispose of certain waste in country B, it must communicate first. This applies to the import, export and even transit of waste, in addition to respecting the sovereignty of each country's laws regarding the classification of waste as hazardous. Thus, we have the creation of international trade, which is often accepted by poorer and marginalized countries, certainly making a profit.

Brazil joined the Basel Convention through Decree No. 875, of July 19, 1993, but it was Decree No. 4.581, of January 27, 2003, which annexed the types of waste that are hazardous and can be controlled. That said, in Annex VIII of Decree No. 4.581, in List A, which mentions the chemical elements that are considered hazardous, electronic components are mentioned.

> A1180 Waste or scrap electrical or electronic assemblies[2] containing components such as accumulators and other batteries included on list A, mercury switches, glass from cathode-ray tubes and other activated glass and PCB capacitors, or contaminated with Annex I elements (e.g. cadmium, mercury, lead, polychlorinated biphenyl) to the extent that they acquire any of the characteristics contained in Annex III (note the corresponding item on list B - B1110)[3]

Although this decree was a way for Brazil to legalize the import of hazardous waste or prevent its disposal in Brazilian territory, Federal Law No. 12.305, of August 2, 2010, which establishes the National Solid Waste Policy (PNRS), states in Art. 49 that "the import of hazardous solid waste and tailings, as well as solid waste whose characteristics cause harm to the environment, public and animal health and plant health, is prohibited, even for treatment, reform, reuse and disposal". 49 that "the importation of hazardous solid waste and waste, as well as solid waste whose characteristics cause damage to the environment, public and animal health and plant health, is prohibited, even if for treatment, reform, reuse, reuse or recovery" (Brasil, 2010, p. 17). In view of this, the table below shows some of the chemical components found in technological waste and which are characterized as hazardous by Decree No. 4.581/2003.

Table 3 - Chemical components present in e-waste

Some chemical components present in e-waste	
Components	Materials
Lead	solder in printed circuits, cathode ray tubes in monitors and televisions
Cadmium	resistors, infrared detectors and semiconductors
Mercury	batteries, household switches and printed circuit boards, monitor
Arsenic	integrated circuits, semiconductors, cell phones, solar cells
Beryllium	heat sinks, cell phones, etc,

Source: Gonçalves (2007 apud Mattos; Mattos; Perales, 2008, p. 5-7)

On the other hand, we can't refrain from introducing the provider of human subsistence, housing, storage, regulator and inhibitor of human waste. It is full of possibilities, functions, has various types and holds within itself various resources, it can also be a bit shapeless, dry, with few nutrients, that said, here we make a reference to the soil. Thus, according to Embrapa (2018, p. 27) soil:

> is a collection of natural bodies, consisting of solid, liquid and gaseous parts, three-dimensional, dynamic, formed by mineral and organic materials that occupy most of the surface mantle of the continental extensions of our planet, contain living matter and can be vegetated in the nature where they occur and, eventually, have been modified by anthropogenic interference.

In the same vein, Mattos, Mattos and Perales (2018) say that certain equipment, when destroyed or even exposed to the ground, can release toxic substances such as mercury and contaminate it, as well as, depending on the chemical component, contaminate river deposits. Furthermore, in the case of informal e-waste management, the burning process can lead to the release of toxic substances into the air, as well as their leaching process.

Still on the subject of e-waste contamination, the report prepared by the World Health Organization (WHO), entitled Children and digital waste: exposure to e-waste and children's health, raises issues related to the Basel Convention, but in a human context. The WHO (2022) refers to e-waste as a serious problem in the poorest countries, territories where some of the inhabitants live close to technological waste dumps.

Certainly, the reality behind WEEE relates to what was said at the beginning of this chapter, the beneficial side of technological expansion, the reduction of space-time. However, beyond the benefits, the improper disposal of electronics brings us back to the inhabitants, especially children, of countries known as e-waste dumps, who, according to WHO (2022), are involved in the manual sorting of electronic components because they have small hands. In addition, according to the Global E-waste Monitor (2022), improper (informal) recycling of e-waste leads to problems such as stress, headaches, shortness of breath, chest pain, weakness and dizziness.

Therefore, we bring this other perspective to this work because, with the logic of consumerism, we are removed from reality and live in a spectacle (a bubble) full of images that imprison us in the consumerist and marketing ideology of capitalism and we can't see the product beyond use, that is, pleasure.

Referring to the spectacle, Debord (1967, p. 13) says that "the life of societies in which modern conditions of production reign is enunciated as an immense accumulation of spectacles", so the spectacle, "in all its particular forms of information or propaganda, advertising or direct consumption of entertainment, constitutes the model of socially dominant life". So we see, as Bodart (2020, n.p.) states, that a:

> he modern society is dominated by a "spectacle", i.e. a set of images and messages that are conveyed by the media and the market and which aim to promote consumption and conformity. These images and messages form a fictitious world that people tend to confuse with reality, which leads to alienation and a loss of critical awareness.

Finally, consumerism is influenced by the mediatization of merchandise, which means that not only do products, such as electronics, have a reduced lifespan, but perception is also bought, taken over by colors, words, inferiority and alienation of consumers.

On the other hand, with regard to e-waste, we have the construction of a vicious cycle, submissive to having, enjoying and exchanging, rejecting producing and discarding. In this sense, there is a web of hidden relationships behind every device we use, in other words, we know that a smartphone, for example, has in its production chain stages involving the extraction of raw materials, the construction of the component, assembly, among others, however, within each stage there is distribution (a logistics) and this extends to other countries, we add, mentioning scientific management, that one of Taylor's ideas, table 1, principle 2, was to separate mental and manual labor, in this way, the production of the cell phone is restricted to two groups: the head office (system developers, programmers and T.I) and factories (dispersed production in countries where labor is cheap, they just assemble). So, in addition to many peripheral countries being garbage dumps, they are also providers of cheap labor. Finally, this relationship of extraction, elaboration and production is intertwined with power players (politicians, businessmen, etc.), who seek to take advantage of the problem in order to profit.

CHAPTER 3: Legislation and reverse logistics of e-waste in Brazil

As explained above, e-waste can be defined as electronic equipment that no longer works and should therefore be deposited at a collection point for proper disposal. That said, there is legislation in force both in Brazil and in Campos dos Goytacazes - RJ on the management of technological waste. Therefore, this chapter will be based on SINIR, Federal Law 12.305 of August 2, 2010, which establishes the National Solid Waste Policy and CONAMA Resolution 401/2008.

3.1 Overview of e-waste in Brazil: legislation and reverse logistics

In the previous chapter, we talked a little about the problems that electronic waste can cause when it is disposed of improperly on the ground. We also looked at the Basel Convention and tried to follow a path from machines, from the first Industrial Revolution, to the mid-20th century, relating the advance of technology and science and their influence on the development of increasingly agile and portable technologies. Thus, from the 1990s onwards, globalization began to integrate the world, albeit unevenly, leading to an increase in world consumption and, consequently, an increase in the amount of waste produced, which often ends up being disposed of in poorer countries.

As far as Brazil is concerned, Federal Law No. 12.305 of August 2, 2010, which establishes the National Solid Waste Policy (PNRS), is the foundation and the path for the correct management of solid waste in Brazilian territory. Beforehand, solid waste is classified in Article 3, item XVI of Law 12.305/2010 as:

> discarded material, substance, object or good resulting from human activities in society, to whose final destination one proceeds, proposes to proceed or is obliged to proceed, in solid or semi-solid states, as well as gases contained in containers and liquids whose particularities make their discharge into the public sewage system or into bodies of water unfeasible, or require solutions that are technically or economically unfeasible in view of the best available technology (Brazil, 2010, p. 2)

E-waste, therefore, can be considered as household waste when used by the population. However, unlike plastic and cardboard (which are the most popular in the media and have large cooperatives recycling them), electronic waste cannot be disposed of, with the exception of some types of batteries, in the common garbage, and therefore needs a specific and well-organized place (ventilated and covered) for disposal. We've already told you that electronic waste goes through reverse logistics, which is a..:

> an economic and social development instrument characterized by a set of actions, procedures and means designed to enable the collection and return of solid waste to the business sector, for reuse in its cycle or in other production cycles, or other environmentally appropriate final destination (Brazil, 2010, p. 2).

Thus, Art. 33 of Law 12.305 states that:

> manufacturers, importers, distributors and traders are obliged to structure and implement reverse logistics systems, by returning products after use by the consumer, independently of the public urban cleaning and solid waste management service (Brazil, 2010, p. 13).

In addition to the above, we would like to point out that Decree No. 10.240, of February 12, 2020, now regulates item VI (electrical and electronic products and their components) of the **main section of** art. 33 of Law No. 12.305/2010. With this regulation, we have, in the aforementioned decree, the definition of waste equipment

The list of products that can be used for reverse logistics includes Annex I, which contains a list of electrical and electronic products that can be used for reverse logistics. A table with some of the products that can go through reverse logistics will be explained below, but we would like to point out that there is technological waste that follows its own reverse logistics, such as hospital waste, corporate waste, batteries, among others.

Looking at table 4 below, we can see some everyday products that we use and which, when they stop working, we often accumulate, and if we are not aware of the collection points, in the worst cases, this can lead to them being disposed of in the general waste without worrying about possible contamination.

Electronic products subject to reverse logistics	
Shaver	Portable air conditioner
Epilator and hair trimmer	Vacuum cleaner
Television set	Video games
Radio set	Babysitting
External battery	Cell phones
Scales	Computers
Mixer	Electric toothbrush
Refrigerated Drinking Fountain	Stoves
Ink or toner cartridge	iron
Electric shower	washing machine
CPU - Central Process Unit	microwave

That said, the figures below, according to The Global E-waste Monitor 2020, represent four e-waste management scenarios. Thus, figure 7 represents the scenario in which e-waste is formally collected, i.e. current legislation is applied, then there is integrated collection, which involves municipal e-waste collection points and stores that receive electronic products for reverse logistics (e.g. cell phone operators that receive obsolete cell phones and send them for recycling). Finally, once collected, the electronics will be sent to a specialized facility that will process them, separate the valuable components, control the hazardous substances and those materials that cannot be reused will be sent to landfill or have another sustainable purpose.

Figure 7 - First scenario

Source: The Global E-waste Monitor (2020, p. 38).

Below, figure 8 highlights a collection related to the disposal of e-waste in garbage cans with ordinary waste such as plastics, paper, food, etc. The consequence of this type of disposal is that e-waste will be considered regular mixed household waste. In this context, there is the possibility of them being incinerated or landfilled without any kind of treatment, we stress that this type of treatment is not appropriate because it can cause problems for the environment and the loss of resources.

Figure 8 - Second scenario

Source: The Global E-waste (2020, p. 39).

Figure 9 below depicts the treatment of e-waste based on collection by individual resellers or companies, who may also be traders in other channels. Therefore, once collected, waste electrical and electronic equipment follows two paths, the first relates to the separation of plastics and metals and directing them to recycling, the second, these materials have the possibility of being exported, therefore, this type of collection can lead to the treatment of these materials in non-specialized locations.

Figure 9 - Third scenario

Source: The Global E-waste Monitor (2020, p.40)

Figure 10 shows a scenario related to informal work, in which products are collected or bought at the door of homes, businesses and other institutions. Once the product has been collected or dealt with informally, the person who bought it can sell it to be repaired, dismantled or recycled. In the case of recycling, depending on the recyclers, the WEEE can be burned and the dismantlers separate out the valuable parts to be used for sale or another profitable function.

Figure 10 - Fourth scenario

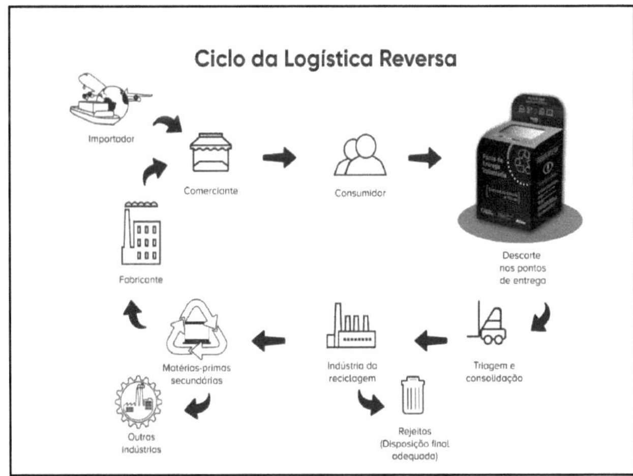

Source: The Global E-waste Monitor (2020, p. 41).

Returning to the first scenario, the figure below shows the structure of reverse logistics.

Figure 11 - Reverse logistics of e-waste

Source: SINIR(2024)

Thus, reverse logistics can be explained as the consumer disposing of their electronic material at collection points, which will then be sent for sorting and the next step is the recycling industry, which will separate the components that should be sent for final disposal (landfill) and the secondary raw materials that will be sent to other industries or manufacturers. The main objective of reverse logistics for e-waste is to reintroduce it to industry, without most of its components being wasted, which in turn avoids the generation of new components and promotes savings in industrial production.

To conclude this first part, we would like to clarify that batteries follow a different reverse logistics from electronic products, according to Law 12.305/2010. That said, according to SINIR (2024), batteries are electrochemical devices that convert chemical energy into electricity, so when they are disposed of inappropriately, they lead to soil and water contamination, and because they contain heavy metals such as lead, mercury and nickel, they cause various problems such as kidney and nervous system problems. Figure 12 has some differences when compared to figure (11) mentioned above.

Figure 12 - Reverse logistics for batteries

Source: SINIR (2024)

The main difference lies in the recycling industry, which will obtain the raw materials for later forwarding to factories or other industries. Unlike the reverse logistics of electronic products, the logistics of batteries adds two more stages after the recycling industry: the separation of metallic zinc (grayish powder generated with functions for other industries in the production of jewelry, diamond tools, etc.) and salts (used in hair dyes) and metallic oxides (used in the coloring of ceramic glazes).

Having explained the differences in the type of logistics, we should make one more observation regarding batteries, which is based on CONAMA resolution 401/2008, which establishes the maximum limits of lead, cadmium and mercury for batteries marketed in the national territory and the criteria and standards for their environmentally appropriate management, and makes other provisions. This resolution, in Art. 22, highlights the prohibition on disposing of any type of batteries in the open air, in urban or rural areas, open burning or incineration without licensed equipment, and finally prohibits disposal in bodies of water. We would also point out that Article 7, which specifies zinc-manganese and alkaline-manganese batteries, leaves open the disposal of these batteries in landfills, which differs from Articles 10 (lead-acid batteries) and 13 (nickel-cadmium and mercury oxide batteries), which prohibit the incineration and disposal of these types of batteries in any type of landfill.

In general, collection points must comply with Resolution 401/2008, because unlike cell phones (without the battery), printers and video games, batteries can cause explosions (small or large) and hurt the people sorting them. Finally, the Manager for National Waste Electrical and Electronic Equipment - Green Eletron, said that in 2022, Brazil had 8,000 voluntary delivery points (PEVs) and 141 tons of batteries had been disposed of correctly.

CHAPTER 4: The dynamics of e-waste disposal in the municipality of Campos dos Goytacazes

Campos dos Goytacazes is a municipality in the state of Rio de Janeiro, located in the northern region of the state of Rio de Janeiro. The author of this research, as explained in the introduction, became interested in e-waste because it is a topic that many people treat as something innovative, but they don't understand that there is a system behind every technology, i.e. a cell phone doesn't assemble itself, it doesn't recycle itself, it doesn't sell itself, so the term innovative must be rejected, because this research seeks to find ways beyond cliché phrases such as: "We have to recycle, if we don't, the world will perish", "We have to become good friends of mother Earth", so, far from what has been said, there is no hierarchy, which plastic pollutes more than e-waste, glass is more dangerous and cardboard less, we emphasize that each material has its specificity and if it is not disposed of correctly, it will become harmful to people, flora and fauna, or, on the other hand, they will be a source of profit for certain groups that get rich from environmental problems.

Thus, once the research object had been identified and the space had been cut out, we searched for laws that were related to the context of e-waste. Successfully, we obtained Federal Law No. 12.305, of August 2, 2010, which paved the way for knowledge of Municipal Law No. 8232, of June 15, 2011 and the Municipal Plan for Integrated Solid Waste Management (PMGIRS), which is still subject to public hearings in the first half of 2024. Campos dos Goytacazes, compared to other municipalities, has implemented the law very slowly. With regard to the word electronic waste, it was not mentioned once, but Article 15 (special waste and refuse) of Law 8232/2011 mentions batteries, which are special waste. In relation to reverse logistics, it is mentioned three times, but we make a very important observation when we refer to Art. 30 of the law establishing the national solid waste policy, which states:

> Shared responsibility for the life cycle of products is instituted, to be implemented in an individualized and chained manner, covering manufacturers, importers, distributors and traders, consumers and holders of public urban cleaning and solid waste management services (Brasil, 2010, p. 12).

Parallel to Brazil (2010, p. 12), Law 8232/2011 expresses the following in Art. 14, item 2:

> The generator of household and similar waste ceases to be responsible when it is properly made available for collection or, in the case of reverse logistics, when it is returned to the places designated for this purpose (Campos, 2011, p.6).

If we compare the two quotes, we can see that the latter law removes the responsibility for the correct disposal of household waste from the generators for collection and reverse logistics, which opens the door to disposal in inappropriate areas.

After accessing the federal law and, consequently, the municipal law, a search was carried out on the Campos dos Goytacazes City Hall website - https://www.campos.rj.gov.br/. Thus, in the first part of the fieldwork carried out in September 2023, the collection points found, as shown in the map in figure 13, were the Municipal Garden (1) and the Environmental Education Center (2).

Figure 13 - Location of collection points

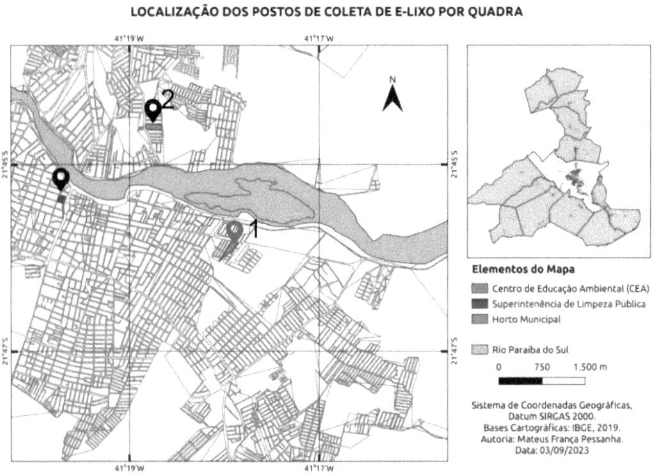

Source: author

The Municipal Garden is located on Avenida Alberto Lamego, next to the State University of Northern Rio de Janeiro Darcy Ribeiro (UENF). This was the first place we visited. From the outset, it is a place that has the correct environment for storing e-waste. In addition, we can see two barrels used to store e-waste, however, as can be seen in figure 14, the place, as of the date of the photo (September 2023), is practically empty. It can therefore be seen that the low level of disposal at this collection point may be related to the Covid-19 pandemic.

Figure 14 - Horto Municipal collection point

Source: authors (September 2023)

On the other hand, with regard to the amount of electronic waste disposed of at the Horto, it was not possible to obtain it, except for one file, which shows the month of August 2022 with 15 kg of electronic waste received. In addition, due to the low level of disposal, until September 2023 the collection was done by individuals.

Further on, the Environmental Education Center (CEA), another collection point, is located on Av. José Carlos Pereira Pinto, in the subdistrict of Guarus. A larger amount of e-waste was found at this point, but unlike the first point, at the CEA the collection is carried out by Caparaó Reciclagem (a company that is called in). As we can see, the waste is kept in barrels, which, once they have been filled and Caparaó has been notified, will be collected. At this collection point, the barrel is close to the service area and if there are any batteries in it, there could be an incident. That's why, when we brought up the laws, it was to highlight the importance of having an appropriate place to store the waste and then send it for sorting and, consequently, reverse logistics.

Figure 15 - CEA collection point

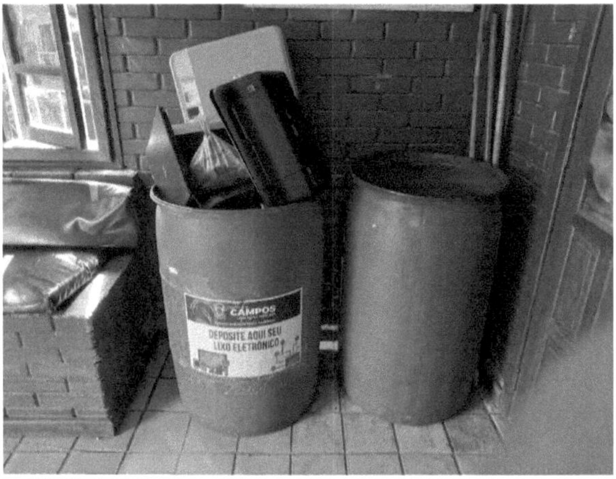

Source: authors (October 2023).

Caparaó Reciclagem is the municipality's main company when it comes to recycling and creating campaigns about electronic waste. It sorts the plastic, glass and other items that are bundled with the electronics. That said, the company said that once the e-waste has been sorted, it is sent to Lorene, a company in São Paulo that carries out reverse logistics. The pictures below, 16 and 17, show the area where the technological waste is sorted.

Figure 16- E-waste in Caparaó

Source: authors (November 2023).

Figure 17 - Secondary e-waste at Caparaó

Source: authors (November 2023).

Figures 16 and 17 therefore show some of the sorting of electronic equipment waste. In addition, figure 16 already shows the electronic waste that will be sent to Lorene, and we would add that this sorting is done manually, there is no burning process, only dismantling. Figure 17 shows the part that will be sent for recycling or to the landfill.

Therefore, when asked for the amount of electronic equipment waste collected, we were informed that Caparaó sent an e-mail reporting on the change in the way the amount of e-waste sent for reverse logistics is evaluated. In addition to the above, the company pointed out that it has a monthly average of 500 kg of e-waste delivered to the final destination, and also said that it would be owed data on the amount of e-waste purchased and the other parts dismembered from it, such as plastic, destined for the landfill.

Therefore, knowing that all the municipal e-waste collection points found were Horto Municipal, CEA, Estande do Meio Ambiente (located in Farol), Superintendência de Limpeza Pública (SULIMP), we add Green Recicla Pilhas, a program run by the Gestora para Logística Reversa de Equipamentos Eletroeletrônicos (Green Eletron), a company created by the Associação Brasileira da Indústria Elétrica e Eletrônica (ABINEE) (Brazilian Association of the Electrical and Electronic Industry) to support companies with regard to Law No. 12.305/2010. As a result of the Green Recicla Pilhas program, in addition to the municipal collection points, twenty-six battery collection points were found in the municipality of Campos dos Goytacazes, concentrated in pharmacies, stationery stores, appliance stores and retail chains.

After all this, as we were unable to obtain quantitative data on the exact amount of e-waste disposed of in a given period of time (year), we tried to apply mixed questionnaires, which allow for greater freedom of response, to people living in the district where the municipality is located - Campos dos Goytacazes. In this way, we applied snowball sampling, considered by Vinuto (2014, p. 203) to the questionnaire prepared and distributed by Google Forms:

> Snowball sampling is a form of non-probability sampling that uses reference chains. In other words, from this specific type of sampling it is not possible to determine the probability of selecting each participant in the survey, but it is useful for studying certain groups that are difficult to access.

In addition to the above, Vinuto (2014) says that snowball sampling starts from the idea of key informants, who will be the basis for locating people with the research profile and, after that, the (key) people selected first will indicate other people from their personal network, thus expanding the application of the questionnaires.

As far as this survey is concerned, the questionnaires were administered between January 2024 and February 2024, and we received a total of 22 responses. Below are some graphs highlighting some of the important questions on the form.

Graph 1 below shows whether the participants in the survey knew what e-waste was. All of them answered yes, which is a good thing, because just by knowing what e-waste is, the awareness-raising part becomes achievable without too many problems.

Graph 1 - Question about e-waste

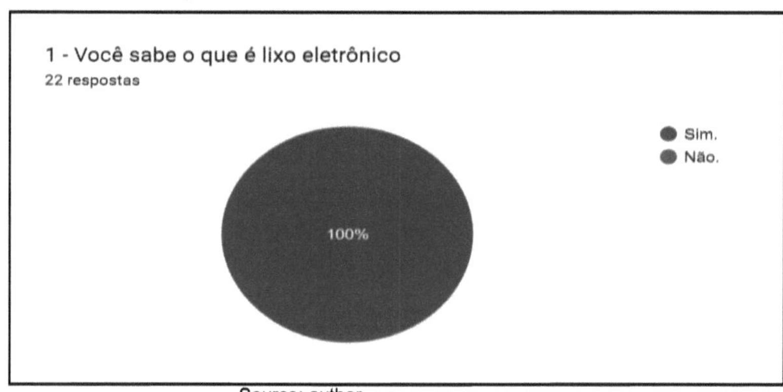

Source: author

Comparing the answers in graph 2 (on the durability of electronic products) and graph 3 (changing cell phones between 2015 and 2023), both below, we can see that, in graph 2, 62.2% said that the lifespan of electronic products had decreased over the years and 31.8% said that they lasted less, thus observing and confirming the existence of programmed obsolescence, as explained in the previous chapter. In addition, we have added graph 3, in which the majority of responses were related to changing cell phones more than three times in an eight-year period, which would give us, for example, in the context of three cell phones, a lifespan of approximately two and a half years for each of them.

Graph 2 - Durability of electronics

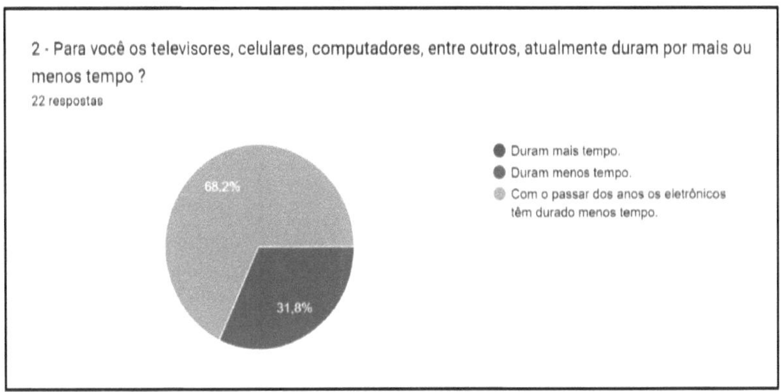

Source: author

Graph 3 - Switching cell phones between 2015 and 2023.

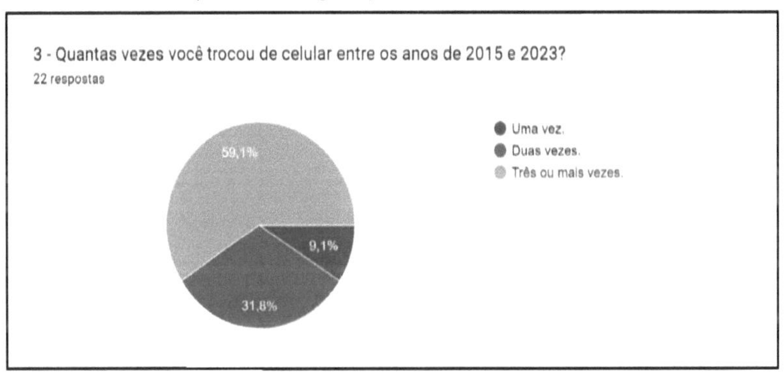

Source: author

Graphs 4 and 5 highlight the disposal of e-waste in the regular garbage, as shown in both graphs, 77.3% have already disposed of their electronic waste in the regular garbage and more than 5% sometimes dispose of it. This brings us to graph 6, in which 95.5% have no e-waste collection near where they live. We see that 4.5% sometimes dispose of batteries in the general waste and 9.1% sometimes dispose of electronic devices in the general waste. Therefore, if we compare the two graphs, we note that there was no response related to correct disposal, which leads us to a problem found and highlighted in this chapter, that is, there are few collection points and the correct dissemination of their existence is concentrated on social networks and city hall websites. This problem also extends to Green Recicla Pilhas, which is publicized on a specific website.

Graph 4 - Disposal of batteries in normal waste

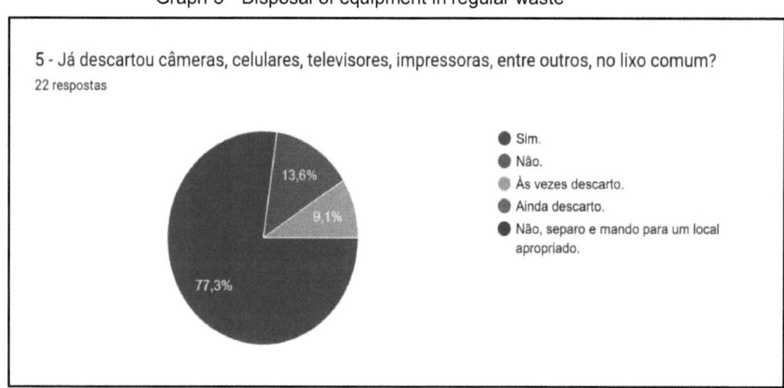

Source: author.

Graph 5 - Disposal of equipment in regular waste

Source: author

The following graph shows that 95.5% do not have an e-waste collection point near where they live. This result was to be expected given that e-waste collection points are located in central areas, which often have to be accessed by car or public transport.

Graph 6 - Existence of a collection point near the address

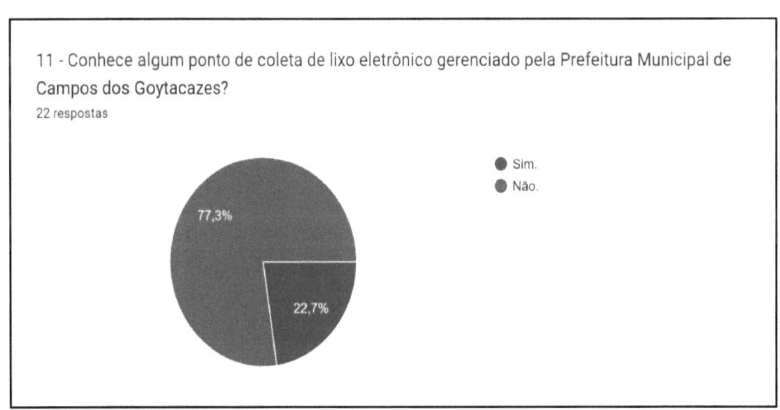

Source: author.

Finally, following Graphs 7 and 8, we would add that the majority of those who answered the questionnaire were also unaware of the existence of technological waste collection points managed by Campos dos Goytacazes City Hall, as well as their understanding of reverse logistics. One possible cause of this lack of knowledge is related to the information, as the municipality publishes it on its website or social networks, making it impossible for people who are not technologically savvy or interested to access it.

Graph 7 - Knowledge of collection points in Campos dos Goytacazes

Source: Author

Graph 8 - Reverse logistics

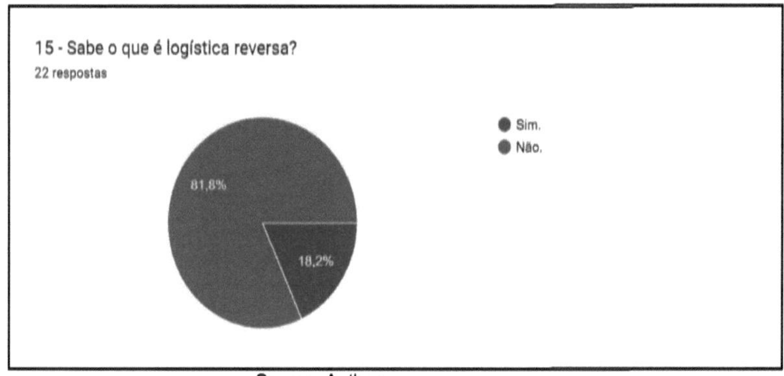

Source: Author

Following this overview of the questionnaires applied, we would like to highlight the dynamics of e-waste in the municipality of Campos dos Goytacazes, which can be seen in the flowchart below, in figure 18. Beforehand, we should add that Campos has selective collection, which sends recyclable materials to recycling cooperatives and materials that cannot be recycled are sent to the landfill. With regard to electronic waste, batteries and electronic products are disposed of at collection points, but Green Recicla Pilhas must be taken into account, in which case some private battery collection points are attached.

Figure 18 - Results of the dynamics of e-waste disposal and collection in Campos

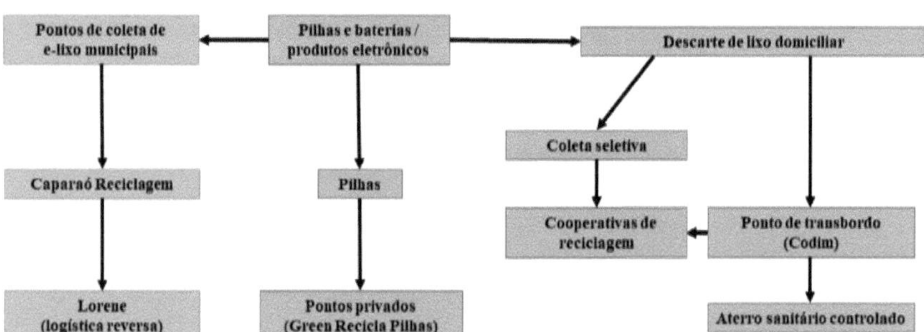

Source: author

All the graphs, despite the fact that the number of questionnaires answered was not that large, demonstrate and corroborate what has been said in this work, i.e. there is an increase in the consumption of electrical and electronic products and their lifespan is being reduced. What's more, as we pointed out in the previous chapter, many people still dispose of e-waste irregularly, which also doesn't exempt the bodies that manage the collection points from their commitment to Law 12.305/2010 and the laws at state and municipal level. Thus, the results of these questionnaires, however simple they may be, expose a problem that exists and must be seen from outside the bubble.

CONCLUSION

We would like to point out that the methodology used in this research, from obtaining websites, laws and questionnaires to international data, was of the utmost importance for its realization. We would also like to add that through it, new paths related to the disposal and collection of electronic waste have been opened up, making it possible to expand the Dynamics of disposal and collection of e-waste in Campos dos Goytacazes to other trajectories such as the stages of reverse logistics of technological waste in practice.

So, in the first part of this research, we looked at the First Industrial Revolution, which promoted the advance of technologies and, consequently, social and economic transformations. This was followed by the Fordist and Toyotist production models, which further deepened technological development in terms of mass production.

Therefore, in relation to the trajectory of electronic products, we can see that it began with the development of machines, because these led to automation and the consequent loss of human labor. As a result, knowledge about machines, in view of technical-scientific advances, made it possible for other machines to emerge with purposes other than production, for example, computers, cell phones, television sets, among others, which make life more flexible for human society.

Electronic waste is neither the best nor the special waste, it's just another environmental problem, like plastic, that needs to be looked at carefully. In the case of the municipality of Campos dos Goytacazes, where the research was carried out, we realized that there are collection points and the correct destination, via, until then, Caparaó Reciclagem, to reverse logistics. As this research consisted of finding out if there are collection points, it opened up avenues for other aspects of e-waste. The paths that emerged were the possible route of electronic waste to the landfill, the paths of technological health waste, which was not covered in this work because it follows a different reverse logistics, as well as the relationship of non-municipal e-waste collection points in the dynamics of disposal and collection of e-waste in Campos. Based on the above, this research is just one way of looking at the importance of recycling and taking an accurate social and economic view of the products we buy, use and discard. Finally, in order to better manage e-waste, a possible solution for expanding collection points would be to place them in churches, schools, central areas of neighborhoods, as some of these places concentrate and are frequented by many people, which would result in better waste management and enable the generation of local income.

REFERENCES

ABRAMIDES, M. B. C.; CABRAL, M. do . C. Regime of flexible accumulation and workers' health. São Paulo: **São Paulo em perspectiva**, v. 17, p. 3-10, 2003. Available at: https://www.scielo.br/j/spp/a/P87NC7ZMqpymgR9t3gBG8yh/?format=pdf&lang=pt. Accessed on: Mar. 16, 2024.

ARRUDA. J. J. de A. **The industrial revolution.** São Paulo: Editora Ática S.A, 1988.

BALDÉ, C.P.; FORTI. V., GRAY, V., KUEHR, R., STEGMANN, P. : The Global E-waste Monitor - 2017, United Nations University (UNU), International Telecommunication Union (ITU) & International Solid Waste Association (ISWA), Bonn/Geneva/Vienna. Available at:https://www.itu.int/en/ITU-D/Climate-Change/Documents/GEM%202017/Global-E-waste%20Monitor%202017%20.pdf. Accessed on: March 18, 2024.

BARROS, J. D' A. Fixes and flows: revisiting a conceptual pair. **Cuadernos de Geografía: Revista Colombiana de Geografía**. [s.l.], v. 29, no. 2, p. 493- 504, 2020. Available at: http://www.scielo.org.co/pdf/rcdg/v29n2/2256-5442-rcdg-29-02-493.pdf. Accessed on: 15 Mar. 2024.

BODART, C. What is the society of the spectacle? **Café com Sociologia**, May 17, 2023. Available at: https://cafecomsociologia.com/sociedade-do-espetaculo/. Accessed on: March 17, 2023.

BRAVERMAN, H. **Trabalho e capital monopolista**. Rio de Janeiro: Zahar Editores, 1980.

BRAZIL. Decree No. 875 of July 19, 1993. Promulgates the text of the Convention on the Control of Transboundary Movements of Hazardous Wastes and their Disposal. D.O. 20/07/1993, P. 10050, July 20, 1993. Available at: https://www.planalto.gov.br/ccivil_03/decreto/d0875.htm. Accessed on: 16 Mar. 2024.

BRAZIL. Decree No. 4.581, of January 27, 2003. Promulgates the Amendment to Annex I and Adoption of Annex VIII to the Basel Convention on the Control of Transboundary Movement of Hazardous Wastes and their Deposit. D.O.U of 27/01/2003, p. nº 1, 27 jan. 2003. Available at: https://www.planalto.gov.br/ccivil_03/decreto/2003/d4581.htm. Accessed on: Mar. 16, 2024.

CANÊDO, L. B. **A revolução industrial**. 2º ed. São Paulo: Editora da Universidade Estadual de Campinas, 1986.

CHIAVENATO, I. **Introdução à teoria geral da administração:** uma visão abrangente **da** moderna administração das organizações. 7 ed. rev. and current. Rio de Janeiro: Elsevier, 2003.

CONAMA - National Environment Council. CONAMA Resolution No. 401, of November 4, 2008. Establishes the maximum limits of lead, cadmium and mercury for batteries marketed in the national territory and the criteria and standards for their environmentally appropriate management, and makes other provisions. Available at:https://www.camara.leg.br/proposicoesWeb/prop_mostrarintegra?codteor=145059 1. Accessed on: March 18, 2024.

CUNHA, C. et al. Fetishism and alienation of labor today based on Marx's conceptions. In: XIII Semana de Economia da UESB, 2014, Vitória da Conquista. **Electronic Annals.** [...] Vitória da Conquista: UESB, 2014. p. 1 - 10. Available at: http://www2.uesb.br/eventos/semana_economia/2014/anais-2014/g02.pdf. Accessed on: Mar. 18, 2024.

EMICO, O. Atomic bombs can decimate humanity - Hiroshima and Nagasaki, 70 years ago. **Estudos Avançados,** v. *29*, p. 209-218, 2015.Available at: https://www.revistas.usp.br/eav/article/view/104960. Accessed on: Mar. 18, 2024.

FORTI. V., BALDÉ. C.P., KUEHR. R., BELI G. The Global E-waste Monitor 2020: Quantities, flows and the circular economy potential. United Nations University (UNU)/United Nations Institute for Training and Research (UNITAR) - co-hosted SCYCLE Programme, International Telecommunication Union (ITU) & International Solid Waste Association (ISWA), Bonn/Geneva/Rotterdam. Available at: https://ewastemonitor.info/gem-2020/. Accessed on: Mar. 18, 2024.

GONÇALVES, H., et al. **Brazilian Soil Classification System.** 5. ed., rev. and ampl. Brasília, DF: Embrapa, 2018. Available at: https://www.embrapa.br/busca-de-publicacoes/-/publicacao/1094003/sistema-brasileiro-de-classificacao-de-solos. Accessed on: March 18, 2024.

GREEN ELETRON - Reverse logistics manager. Available at: https://greeneletron.org.br/. Accessed on: Mar. 16, 2024.

HARVEY, D. **Postmodern Condition**. São Paulo: Edições Loyola, 1992.

MAGERA, M. **Os caminhos do lixo**. São Paulo: Átomo Publishing House, 2012.

MARX, Karl; Angels, Friedrich. **History, nature, work and education.** São Paulo: Editora Expressão Popular, 2020.

OLIVEIRA, P. M. de; SANTOS, F. R. dos. Geographical networks in the age of globalization: some reflections on the urban network in its historicity and theoretical practice. [s.l.], **Revista Formação (Online)**, v. 26, n. 47, p. 3-22, jan./abr. 2019. Available at: https://revista.fct.unesp.br/index.php/formacao/article/view/5711/4891. Accessed on: Mar. 17, 2024.

WHO - World Health Organization. Children and digital waste: exposure to electronic waste and children's health. Geneva: World Health Organization, 2022. License:CC BY-NC-SA 3.0 IGO. Available at:

https://www.who.int/pt/publications/i/item/9789240024557. Accessed on: 16 Mar. 2024.

UN - United Nations Organization. Basel Convention. Available at: https://news.un.org/pt/tags/convencao-da-basileia. Accessed on: 02 Apr. 2024.

PASQUINI, N. C. The industrial revolutions: a conceptual approach. São Paulo, **Fatec Americana**, v. 8, n. 1, p. 29 - 44, 2020. Available at: https://www.fatec.edu.br/revista/index.php/RTecFatecAM/article/view/235/206. Accessed on: Mar. 16, 2024.

RIBEIRO, A. de F. Taylorism, Fordism and Toyotism. **Lutas Sociais,** São Paulo, v. 19, n. 35, p.65-79, jul./dez. 2015. Available at: https://revistas.pucsp.br/index.php/ls/article/view/26678. Accessed on: Mar. 16, 2024.

SANTOS, M. **Por uma outra globalização**: do pensamento único à consciência universal. Rio de Janeiro: Editora Record, 2001.

SANTOS, M. **Tecnica, espaco, tempo: globalizacao e meio tecnico-cientifico informacional**. São Paulo: Hucitec, 1996.

SINIR - National solid waste management information system. Reverse logistics. Available at:https://sinir.gov.br/perfis/logistica-reversa/. Accessed on: 17 Mar. 2024.

DISPOSAL OF ELECTRONIC WASTE IN THE MUNICIPALITY OF CAMPOS DOS GOYTACAZES

How old are you?

2 - Do you think televisions, cell phones, computers, etc., last longer or shorter these days?

O - They last longer.

O - They last less time.

O - Over the years, electronics have lasted less time.

3 - How many times did you change your cell phone between 2015 and 2023?

O - Once.

O: Twice.

O - Three or more times.

4 - Have you ever disposed of a battery (cell phone or notebook) or battery in the regular trash?

O: Yes.

O: No.

O - Sometimes I discard it.

O - I still don't think so.

O - No, I separate them and send them to an appropriate place.

5 - Have you ever thrown cameras, cell phones, televisions, printers, among others, in the regular trash?

O: Yes.

O: No.

O - Sometimes I discard it.

O - I still don't think so.

O - No, I separate them and send them to an appropriate place.

7 - Have you ever disposed of cameras, cell phones, televisions, printers, among others, in vacant lots or on the street?

O: Yes.

O: No.

O - Sometimes I discard it.

O - I still don't think so.

8 - Is there any collection of electronic waste where you live?

O: Yes.

O: No.

9 - If the previous answer (8) was "YES", tell us who collects this electronic waste.

10 - If you know of an e-waste collection point, is it accessible (easy to go to and deposit e-waste)?

O: Yes.

O: No.

O - I don't know where the collection points are.

11 - Do you know of any e-waste collection points managed by Campos dos Goytacazes City Hall?

O: Yes.

O: No.

12 - If the previous answer (11) was "YES", write the name of the collection point(s).

13 - Have you heard or read about any campaign by the Campos dos Goytacazes City Council about the importance of disposing of electronic waste in the correct places, such as the Municipal Garden and the Environmental Education Center?

O: Yes.

O: No.

14 - Do you know what programmed obsolescence is?

O: Yes.

O: No.

15 - Do you know what reverse logistics is?

O: Yes.

O: No.

ANNEX B - Table of private collection points for batteries.

Collection point	Address	Neighborhood	City	State
102 - ATACADÃO CAMPOS DOS GOYTACAZES	Avenida Carlos Alberto Chebabe KM 04, S/N	Guarus Park	FIELDS OF GOYTACAZES	RJ
292 - ATACADÃO CAMPO DOS GOYTACAZES	Avenida Doutor Nilo Peçanha, 1038	Santo Amaro Park	FIELDS OF GOYTACAZES	RJ
ASSAÍ 188	Avenida Doutor Nilo Peçanha, 479	Santo Amaro Park	FIELDS OF GOYTACAZES	RJ
CASAS BAHIA CAMPO DOS GOYTACAZES - CENTER	JOÃO PESSOA STREET, 197	Center	FIELDS OF GOYTACAZES	RJ
DROGA RAIA - CAMPO DOS GOYTACAZES B	AV. PELINCA, 231/233, 0	PRQ TAMANDARÉ	FIELDS OF GOYTACAZES	RJ
DROGA RAIA - CAMPOS GOYTACAZES C	DR. LACERDA SOBRINHO, 281, 0	CENTER	FIELDS OF GOYTACAZES	RJ
DROGA RAIA - CAMPOS GOYTACAZES D	AV. TWENTY-EIGHT MARCH, 111, 0	CENTER	FIELDS OF GOYTACAZES	RJ
DROGA RAIA - CAMPOS GOYTACAZES AND	AV. ALBERTO LAMEGO, 1087, 0	PRQ CALIFORNIA	FIELDS OF GOYTACAZES	RJ
DROGA RAIA - CAMPOS	RUA TENENTE	CENTER	FIELDS OF GOYTACAZE	RJ

GOYTACAZES F (LAGOA DOURADA)	CORONEL CARDOSO, 835, 0		S	
DROGA RAIA - CAMPOS GOYTACAZES G (ALVARO TAMENGA)	AV. ALBERTO TORRES, 424, 0	CENTER	FIELDS OF GOYTACAZES	RJ
DROGA RAIA - CAMPOS GOYTACAZES H (SALO BRAND)	AV. TANCREDO NEVES, 85, 0	PRQ JD CARIOCA	FIELDS OF GOYTACAZES	RJ
DROGA RAIA - CAMPOS GOYTACAZES I (TARCISIO MIRANDA)	AV 24 DE OUTUBRO , Nº97, .	-	FIELDS OF GOYTACAZES	RJ
Drogarias Pacheco-DP CAMPOS 2	AV. SEPTEMBER SEVENTH, 484, -	CENTER	FIELDS OF GOYTACAZES	RJ
Drogarias Pacheco-DP CAMPOS 3	R LACERDA SOBRINHO, 12, -	CENTER	FIELDS OF GOYTACAZES	RJ
Drogarias Pacheco-DP CAMPOS 4	BLV FRANCISCO DE P CARNEIRO, 19 - LJ E SOBRELJ, -	CENTER	FIELDS OF GOYTACAZES	RJ
Drogarias Pacheco-DP CAMPOS 5	PC SAO SALVADOR, 35, -	CENTER	FIELDS OF GOYTACAZES	RJ
Drogarias Pacheco-DP CAMPOS 6	AV. TENENTE CORONEL CARDOSO , 683, -	CENTER	FIELDS OF GOYTACAZES	RJ
Drogarias Pacheco-DP CAMPOS 7	R DOCTOR RAUL ABBOTT ESCOBAR, -	CALIFORNIA PARK	FIELDS OF GOYTACAZES	RJ

Drogarias Pacheco-DP PARQUE GUARUS	R SALO BRAND, 58, -	GUARUS PARK	FIELDS OF GOYTACAZES	RJ
Drogarias Pacheco-DP PELINCA	AV. PELINCA, 162, -	PELINCA	FIELDS OF GOYTACAZES	RJ
Drogarias Pacheco-DP PELINCA 2	R ALVARENGA FILHO, 20 24, -	CENTER	FIELDS OF GOYTACAZES	RJ
Drogarias Pacheco-DP PELINCA 3	R VOLUNTARIOS DA PATRIA, 505, -	PELINCA AVENUE PARK	FIELDS OF GOYTACAZES	RJ
Drogarias Pacheco-DP SHOPPING BOULEVARD CAMPOS	R DR SILVIO BASTOS TAV. ARES, 330, LJ 102F 102G 102H 1 PISO, -	LEOPOLDINA PARK	FIELDS OF GOYTACAZES	RJ
KALUNGA - RJ-GOY-SHOP GOYTACAZ	Rua Doutor Sílvio Bastos Tavares, 316, -	Leopoldina Park	FIELDS OF GOYTACAZES	RJ
EXTRA MARKET 3324	Rua Doutor Felipe Uebe, 451/469	California Park	FIELDS OF GOYTACAZES	RJ

Source: https://sistema.gmclog.com.br/info/green

I want morebooks!

Buy your books fast and straightforward online - at one of world's fastest growing online book stores! Environmentally sound due to Print-on-Demand technologies.

Buy your books online at
www.morebooks.shop

Kaufen Sie Ihre Bücher schnell und unkompliziert online – auf einer der am schnellsten wachsenden Buchhandelsplattformen weltweit! Dank Print-On-Demand umwelt- und ressourcenschonend produziert.

Bücher schneller online kaufen
www.morebooks.shop

Printed by Books on Demand GmbH, Norderstedt / Germany